Vitor Chaves de Oliveira

Piezoelectricity in Practice: Prototyping with Embedded Electronics

Project for Clean Energy Generation in Sustainable Urban Infrastructures

ScienciaScripts

Imprint

Any brand names and product names mentioned in this book are subject to trademark, brand or patent protection and are trademarks or registered trademarks of their respective holders. The use of brand names, product names, common names, trade names, product descriptions etc. even without a particular marking in this work is in no way to be construed to mean that such names may be regarded as unrestricted in respect of trademark and brand protection legislation and could thus be used by anyone.

Cover image: www.ingimage.com

This book is a translation from the original published under ISBN 978-613-9-61433-2.

Publisher:
Sciencia Scripts
is a trademark of
Dodo Books Indian Ocean Ltd. and OmniScriptum S.R.L publishing group

120 High Road, East Finchley, London, N2 9ED, United Kingdom
Str. Armeneasca 28/1, office 1, Chisinau MD-2012, Republic of Moldova, Europe
Managing Directors: Ieva Konstantinova, Victoria Ursu
info@omniscriptum.com

Printed at: see last page
ISBN: 978-620-8-61466-9

Vitor Chaves de Oliveira

Piezoelectricity in Practice: Prototyping with Embedded Electronics

Table of contents:

DEDICATORY

I dedicate this work to my parents.

ACKNOWLEDGMENTS

To Prof. Dr. Vagner Vasconcellos
President of the Examining Board and Professor of the Lato Sensu Postgraduate Program - Specialization in Electrical Engineering and Power Systems at UNISAL.
My advisor, for his collaboration and support in the conception of this work.

To Prof. Eng. Sergio Bimbi Junior
For the teachings and discussions on Electronics, Telecommunications, Low Level Programming and for supporting the completion of this course.
Professor at the São Paulo Faculty of Technology (FATEC)

To Prof. Dr. Alexandre de Assis Mota
For the discussions on the subject of Energy Efficiency.
Professor at PUC-Campinas.

To Prof. Dr. Lia Toledo Moreira Mota
For the discussions on the subject of Energy Efficiency.
Professor at PUC-Campinas.

To Prof. MSc. Nivaldo Tadeu Marcusso
Coordinator of UNISAL's Lato Sensu postgraduate courses.

To Prof. MSc. Mauricio Becker
Board Examiner and Professor at PUC-Campinas.

To Prof. Dr. Diogo Gará Caetano
Researcher in Digital Communications Systems at UNICAMP and Coordinator of Lato Sensu Postgraduate Courses at UNISAL.

To Prof. MSc. Lázaro Partamian Carriel
Professor of the Lato Sensu Postgraduate Program - Specialization in Electrical Engineering and Power Systems at UNISAL.

To Prof. Dr. Se Un Ahn
Professor of the Lato Sensu Postgraduate Program - Specialization in Electrical Engineering and Power Systems at UNISAL.

To Prof. Dr. Geraldo Peres Caixeta
Professor of the Lato Sensu Postgraduate Program - Specialization in Electrical Engineering and Power Systems at UNISAL.

To Prof. Esp. Marcelo Rodrigues Soares
Professor of the Lato Sensu Postgraduate Program - Specialization in Electrical Engineering and Power Systems at UNISAL.

To Prof. MSc. Rosivaldo Ferrarezi
Professor of the Lato Sensu Postgraduate Program - Specialization in Electrical Engineering and Power

Systems at UNISAL.

To Prof. Dr. Carlos Alberto Favarin Murari
Professor of the Lato Sensu Postgraduate Program - Specialization in Electrical Engineering and Power Systems at UNISAL.

To Prof. MSc. Juliana Ravaschio Franco de Camargo
Professor of the Lato Sensu Postgraduate Program - Specialization in Electrical Engineering and Power Systems at UNISAL.

To Prof. MSc. Alex Almeida Pignatti
Board Examiner and Professor of the Lato Sensu Postgraduate Program - Specialization in Electrical Engineering and Power Systems at UNISAL.

To my colleagues: Clóvis Vaz Coelho and Paulo Ricardo Polidoro for their encouragement and collaboration in my studies to complete this course.

To the secretaries and administrative staff for their cooperation in responding to requests.

To UNISAL for encouraging teaching, research and extension by awarding scholarships to students, providing suitable laboratories and allocating qualified teaching staff to the Lato Sensu Postgraduate Program - Specialization in Electrical Engineering and Power Systems.

EPIGRAPH

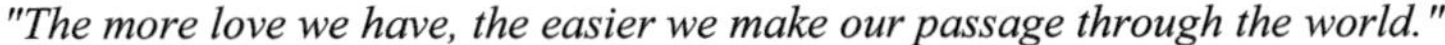

"The more love we have, the easier we make our passage through the world."

Immanuel Kant

SUMMARY

OLIVEIRA, Vitor Chaves de. *Prototype for the Sustainable Generation of Electricity from Motorcycle Losses in Urban Infrastructure through Piezoelectricity*. 2016. Monograph of Course Conclusion Work (TCC) - UNISAL, Centro Universitário Salesiano de São Paulo, Postgraduate Lato Sensu: Specialization in Electrical Engineering and Power Systems.

In recent years, the demand for sustainable electricity production solutions has grown significantly. The reasons for this increase range from indications of a possible exhaustion of the production model based on fossil fuels, to issues of improving air quality in urban environments and even economic concerns. At the same time, people's way of life has become consolidated in the context of cities. Therefore, developing technologies that aim to implement solutions for urban infrastructure systems becomes relevant. Added to this is the fact that electricity is classified as an indispensable commodity for modern human life. And in order to meet this set of needs, sustainable means of generating electricity must be sought. With this in mind, this work has developed a prototype. It uses a technology that can be applied in this scenario and which makes use of energy that is lost to the environment - kinetic energy. And, for the proposed locations, this loss is significantly demonstrated by the mechanization of engines, i.e. motor-kinetics. Piezoelectricity was selected as the technology capable of producing electricity from the losses in this process. Then, to consolidate the necessary combination, the prototype was built using embedded electronics supported by a PIC, Programmable Interface Controller. This type of microcontroller was selected because it has unique features and advantages. In this work, it was programmed in a low-level language, proving to be effective for designing the intended prototyping. As a result, the small-scale prototype is presented, along with an example of its use in the form of a railway model.

Index Terms: Embedded Electronics, Electricity, Sustainable Generation, Urban Infrastructure, Motor-Kinetics, Piezoelectricity, Programmable Interface Controller (PIC).

Chapter 1

1. INTRODUCTION

1.1. Background to the problem

Considering the global electricity context, there has been a continuous growth in demand in recent decades, as shown in Figure 1, with around 13% of this energy being generated from renewable sources and the rest produced primarily from fossil fuels (Statista, 2016 A). It is therefore worth noting that the exhaustion of these sources is expected by the end of this century and is coupled with a possible rise in electricity costs over time if this generation model is perpetuated (CIA, 2016; Ecotricity, 2016). It should also be borne in mind that the cost of electricity has risen over the last 60 years and that there is an indication that its price will rise over time (IMF, 2011; Statista, 2016 B). This upward trend in demand for electricity is closely, but not exclusively, related to another factor that is causing residential consumption to rise, which is the migration of people from rural to urban areas (Oliveira and Simon, 2004). It should be noted that this consolidation of urbanization, i.e. the allocation of people to cities, is a global trend with more than 54% of the population living in these environments (United Nations, 2014; WHO, 2014).

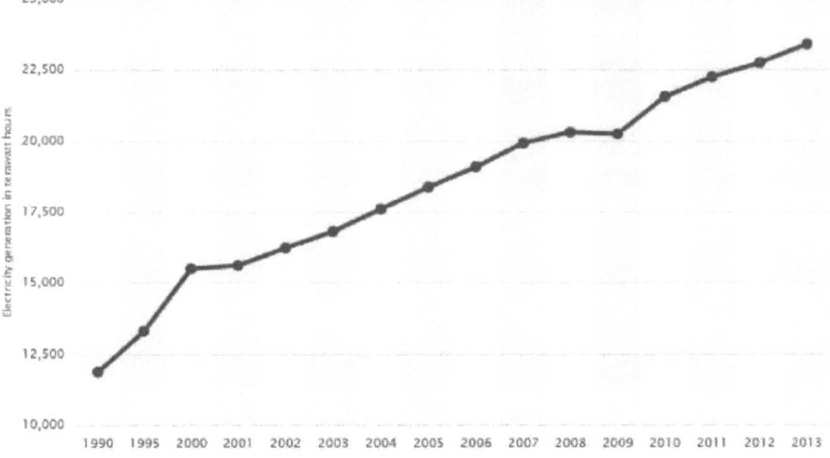

Figure 1: World electricity generation in terawatt hours from 1900 to 2013.
Source: Reproduced and Adapted from Worldwide electricity generation from 1990 to 2013 (in terawatt hours)... (Statista, 2016 A)

In Brazil, these price and urbanization factors have proven to be even more significant and in need of proposals and solutions. This is because Brazil has one of the highest electricity generation costs in the world, as can be seen in Figure 2, which shows that Brazil is the second most expensive country in the world for electricity (Statista, 2016 B). On the question of urbanization, the latest census, shown in Figure 3, shows that the nation is in the last stages of this process, with 84% of the population living in cities (IBGE, 2010).

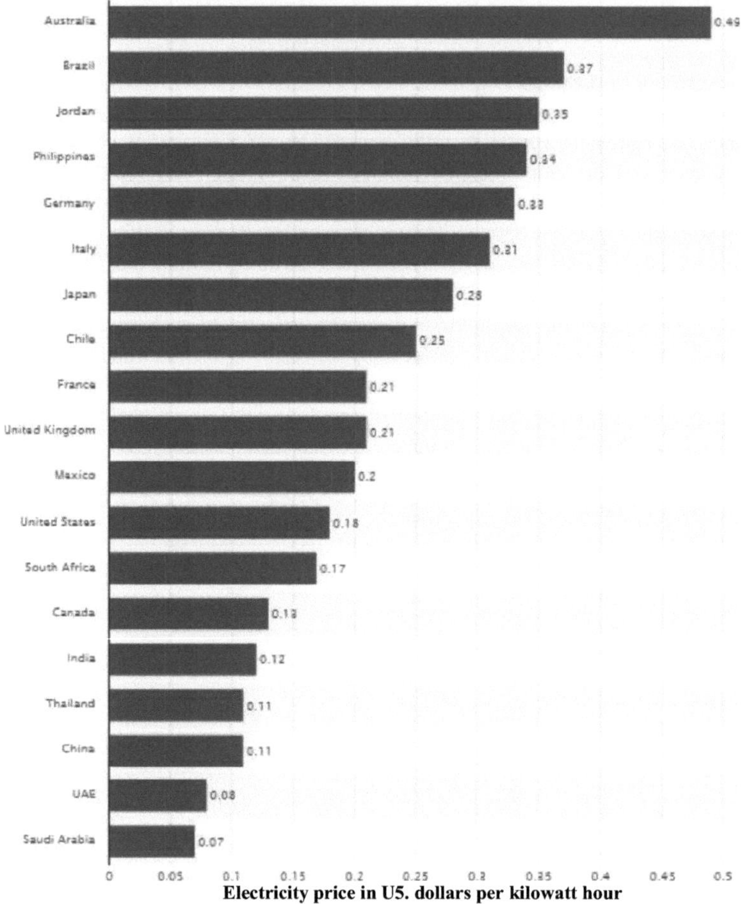

Figure 2: Global average price of a kilowatt hour in 2015 by country, in US dollars.
Source: Reproduced and Adapted from Average electricity prices globally in 2015, by select country (in U.S. dollars per kilowatt hour)... (Statista, 2016 B)

Regi on	Major Regions and Federatio n Units	1960[1] Urban	1960[1] Rural	1970[1] Urban	1970[1] Rural	1980[1] Urban I will order	1980[1] Rural	1M1[2] Urban	1991[2] Rural	2000[2] Urban	2000[2] Rural	201O[2] Urban	year[2] Rural
	BRAZIL	32.004.817	38.987.526	52.904.744	41.603.839	82.013.375	39.137.198	110.875.826	36.041.633	137.755.550	31.835.143	160.925.792	29.830.007
CB	Northern Region	1.041.213	1.888.792	1.784.223	2.404.090	3.398.897	3.368.352	5.931.567	4.325.699	9.002.962	3.890.599	11.664.509	4.199.945
r3	Region North East	7.680.681	14.748.192	11.980.937	16.694.173	17.959.640	17.459.516	25.753.355	16.716.870	32.929.318	14.763.935	38.821.246	14.260.704
0	Southeast Region	17.818.649	13.244.329	29.347.170	10.984.799	43.550.664	9.029.863	55.149.437	7.511.263	65.441.516	6.855.835	74.696.178	5.668.232
EB	Southern Region	4.469.103	7.423.004	7.434.196	9.249.355	12,153.971	7.226.155	16.392.710	5.724.316	20.306.542	4.783,241	23.260.896	4.125.995
E3	Central West Region	995.171	1.683.209	2.358.218	2.271.422	4.950.203	2.053.312	7.648.757	1.763.485	10.075.212	1.541.533	12.482.963	1.575.131

Figure 3: Urban versus rural population in Brazil from 1960 to 2010.
Source: Reproduced and Adapted from Table 1.8 - Population in the Demographic Censuses, by Regions, Federation Units and household situation - 1960/2010... (IBGE, 2010)

Considering the aforementioned trends and needs, activities and studies related to reuse, efficiency and sustainability have become the focus of attention (Canadian Institute of Actuaries, 2015; CIA, 2011). With these challenges in mind, this work set out to develop a hardware and software prototype to generate energy sustainably using piezoelectricity. The reasons for choosing the technologies employed are explained below.

1.2. Justification for developing the work

Considering the growing need for sustainable solutions and the growth of the urbanized environment. And, with this, the increase in the vehicle fleet, especially in metropolitan regions in the country (INCT, 2013), it was realized that a possible gain could be obtained by capturing motor-kinetic energy dispersed into the environment by traffic, that is, by the movement of these cars. And, when looking at the literature, it can be seen that the design of solutions that operate with this objective in these urban infrastructure systems is scarce. In addition, there are few commercial

solutions for generating electricity sustainably using piezoelectricity on a large scale. This highlights the need for further study and development of
proposals around this theme. In other words, developing ways of capturing this energy, which is naturally lost to the environment, and reusing it, thus generating what is known as clean, sustainable energy.

1.3. Objective of the work

In this work, the general objective was to develop a prototype using embedded electronics for the sustainable generation of electricity from motor-kinetic losses in urban infrastructure using piezoelectricity.

To this end, two specific objectives were considered:

1 - Study of the implementation of embedded electronics using PIC16F877 hardware that interacts with electric piezo. Software development with programming in C language and complete simulation of hardware operation using the Proteus tool.

2 - Assembly of the prototype on a phenolite board for electronic circuits by soldering and physically connecting all the components. Installation on a railway model to demonstrate the large-scale use of the technology.

To this end, hardware and software-based methodologies have been developed to apply this technology to urban environments, since these environments have higher motor-kinetic losses.

1.4. Research delimitation

This work is delimited as follows to build the solution to each of the aforementioned specific objectives:

1 - Software programs were built using low-level languages, which have a direct relationship with the hardware, in order to demonstrate the gains possible from varying the pressure in an electric piezo. These programs were recorded on the multiprogrammable hardware of the PIC Platform, as well as being previously simulated in software for testing with the tool Proteus, which allowed a correct study of the solution's behavior (especially interrupts and the state machine) before physical implementation;

2 - On a board with the previously simulated schematic, the PIC was assembled with the programmed instructions, i.e. recorded in the hardware, and all the electrical and electronic

components were connected and soldered, i.e. LEDs, cables, trimpot, power supply and electric piezo. A model railway was built and this board and the piezoelectricity sensor were attached to it to exemplify its application.

It is worth pointing out that this work was limited to building a prototype to demonstrate how the technology works and an example of where it can be applied. It was not the focus of this work to develop studies, calculations and projections for large-scale commercial implementation. This is incipient research, with the aim of demonstrating a technology that is little used in urban infrastructure systems and that is feasible to implement from a technical point of view. This way, such implementations on a larger scale can be based on studies such as the one carried out in this work, which focuses on the prototype.

1.5. Organization of the Monograph

This work is divided as follows.

Chapter 1, Introduction: contextualizes the topic, presents the justification for its study, describes the objective of the work and delimits the actions present in this research.

Chapter 2, Embedded Electronics: describes the operation of devices that allow hardware to be controlled via software: Programmable Interface Controller - CIP or Programmable Interface Controller - PIC.

Chapter 3, Piezoelectricity: discusses the definition of the physical principle behind producing electrical energy, the phenomenon and the devices that use it.

Chapter 4, Methodology: shows how the computer programs were designed and built, the requirements analysis and the detailed development. It also explains the simulation up to the assembly of the electronic circuit, the connections between its components and its operation.

Chapter 5, Results: presents the electronic circuit board built and attached to the train model, exemplifying how it works.

Chapter 6, Conclusion: provides a discussion of the results obtained and the aforementioned implementations, considers the impacts imposed on the subject and outlines possible developments for future work.

Chapter 7, References: lists all the bibliographical references presented in this work.

Chapter 8, ANNEX A: contains the core of the programming coding used in the project in a detailed and commented form.

Chapter 2

2. EMBEDDED ELECTRONICS

Embedded electronics comprises hardware devices that can be programmed via software and attached to a platform, such as a car, an industrial machine, a mobile device, etc. They are generally built to meet a specific need. For this, microcontrollers are generally used, which currently have integrated circuits, microprocessors and are limited to a certain capacity for processing instructions. There are currently several types of microcontroller, among the best known being the PIC (Programmable Interface Controller, CIP), Arduino, Atmel with ARM cores (Acorn RISC Machine or Advanced RISC Machine) and hybrid systems addressing programmable logic and FPGAs (Field Programmable Gate Array). (Costa *et al*, 2011; Miyadaira, 2009; Souza, 2003)

For the purposes of this work, it was found that, for cost-benefit reasons, either the Arduino or the PIC could be used, as both have a low cost for the proposed prototype. And for technical reasons, the PIC was chosen. This is because it outperforms the Arduino due to the fact that its compiler provides the programmer with algorithms with advanced data control and does not require the use of closed libraries that make performance, i.e. the performance of the optimized stack, inferior to data flow control and clock cycle optimization (Costa *et al*, 2011; Miyadaira, 2009; Souza, 2003).

2.1. Programmable Interface Controller - CIP (PIC)

The Programmable Interface Controller or PIC has several types of integrated circuits available on the market, for this work the PIC16F877 was chosen. This is because the PIC microcontrollers of the 16xx family are available on the market with various peripheral architectures available, such as memory capacity and
number of input and output pins, on an 8-bit platform (Costa *et al*, 2011; Miyadaira, 2009; Souza, 2003). This processor was chosen due to its cost and processing capacity being compatible with the project, as well as the possibilities of expanding the system or minimizing it using the same codes, since its hardware registers are in the same family and do not need to be modified if the platform is maintained.

Considering the PIC 16F877, we can see that it is a microcontroller with a 14-bit core and an 8-bit family. This PIC is manufactured by a company called Microchip Technology. The part of the name that contains the number 16 indicates that it is part of the "MID- RANGE" family, being an 8-bit microcontroller. This means that the ALU, or Arithmetic and Logic Unit, operates with a limit of up to 8 bits for each set of 'words' at a time. While the F in this name tells you that the memory that makes it up is "Flash", a type of memory that allows fast access, typified in the

EEPROM category (Electrically Erasable Programmable Read-Only Memory) which allows multiple registers to be modified (written and/or erased) at a time, i.e. in the same operation. In this schematic, it is also worth noting that each 14 bits (including commands and addresses, 8 of which are data bits) represent one line of this memory, i.e. it forms 'words' of this size. In addition, there is the complete and precise identification of the PIC in question provided by the remaining three characters of this name, identified by 877. This specific model has a processing speed of around 20Mhz, however the reference may be accompanied by a -XX suffix, indicating the maximum operating frequency of the clock present in the PIC. All this information can be seen in Figure 4, taken from the PIC Datasheet supplied by the manufacturer. Figure 4 also shows a comparison with other PICs and also shows the larger memory capacities, interrupt handling and input and output ports (Microchip Technology, 2013).

Key Features PIC® MCU Mid-Range Reference Manual (DS33023)	**PIC16F873**	**PIC16F874**	**PIC16F876**	**PIC16F877**
Operating Frequency	DC - 20 MHz	DC - 20 MHz	DC - 20 MHz	DC - 20 MHz
RESETS (and Delays)	POR, BOR (PWRT, OST)	BY. BOR (PWRT, OST)	BY. BOR (PWRT, OST)	BY. BOR (PWRT, OST)
FLASH Program Memory (14-bit words)	4K	4K	8K	8K
Data Memory (bytes)	192	192	368	368
EEPROM Data Memory	128	128	256	256
Interrupts	13	14	13	14
I/O Ports	Ports A.B.C	Ports A,B,C,D,E	Ports A.B.C	Ports A.B.C.D.E
Timers	3	3	3	3
Capture/Compare/PWM Modules	2	2	2	2
Serial Communications	MSSP, USART	MSSP, USART	MSSP. USART	MSSP. USART
Parallel Communications	-	PSP	-	PSP
10-bit Analog-to-Digital Module	5 input channels	8 Input channels	5 input channels	8 input channels
Instruction Set	35 instructions	35 instructions	35 instructions	35 instructions

Figure 4: Comparative table with the main characteristics of PICs - PIC16F87X.
Source: Reproduced and Adapted from Key Features PIC® MCU Mid-Range Reference Manual (DS33023)... (Microchip Technology, 2013)

The architecture of the PIC16F877 is shown in Figure 5 below, where you can see how communication occurs in this circuit, as well as how the components relate logically and physically within this device. This illustration is very useful for understanding how to program the PIC to control hardware devices in general, how to identify the pinout, where to physically receive and send signals, how to read states, as well as receive and process information. It should also be noted that the figure is in the format of a block diagram in which it is possible to see the physical path of the inputs and outputs, thus collaborating with the software design to conceive the project. (Microchip Technology, 2013)

PIC16F87X

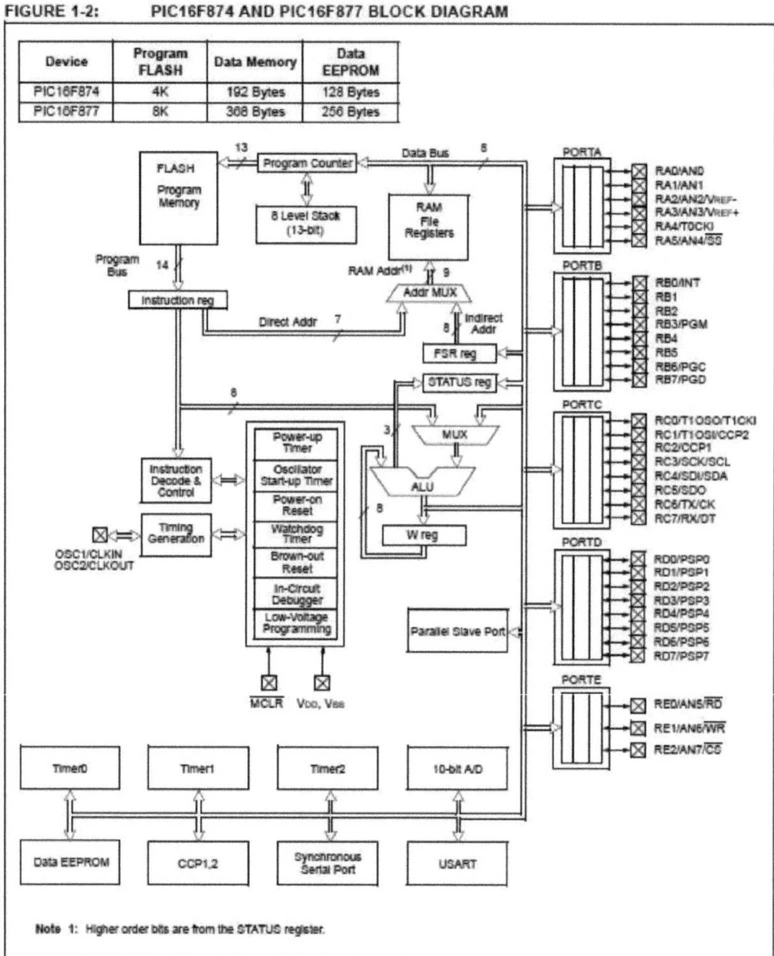

Figure 5: PIC Architecture Breakdown, Block Diagram - PIC16F87X.

Source: Reproduced and Adapted from FIGURE 1-2: PIC16F874 AND PIC16F877 BLOCK DIAGRAM... (Microchip Technology, 2013)

Chapter 3

3. PIEZOELECTRICITY

Piezoelectricity is an electrical charge that accumulates in certain materials when they are subjected to a variation in mechanical pressure, i.e. mechanical stress. This effect is observed in various types of materials such as proteins, DNA, bones, ceramics and crystals. It is interesting to note that the word itself, piezoelectricity, derives both from the Greek 'piezein' which means pressure, and from the also Greek 'electron' which is assigned to the material amber, one of the first in which electricity was observed. The history of piezo electricity shows that this effect was discovered by the French brothers Pierre and Jacques Curie in 1880. This was made possible by crystals and the inverse piezoelectric effect was eventually deduced mathematically by thermodynamic principles by Gabriel Lippmann the following year. Following this, the Curie brothers experimentally confirmed this reversibility of mechanical changes in crystals, which were then characterized as piezoelectric. Throughout these times, various studies have continued in order to categorize whether or not a material possesses this property of generating a voltage or potential difference that translates into the production of an electrical charge, i.e. electrical energy, when a mechanical stress is applied. And in 1910, a German book appeared, the 'Textbook on Crystal Physics'. It established two dozen categorizations of natural crystals segmented by checks of mechanical and tensorial stress analysis, obtaining piezoelectric constants for each of these classes. (Arnau Vives, 2008; Cotta; 2007)

3.1.. Description of operation

The piezoelectric effect, as previously mentioned, is more noticeable in certain crystalline materials, such as ceramics and polymers. It is understood in practical terms as an electromechanical interaction, i.e. a linear relationship between the Coulomb forces of the electrical state, linked to voltage, and the mechanical force, linked to the physical stress of the material due to pressure variation. This effect is a reversible process in materials that demonstrate the so-called direct piezoelectric effect, i.e. both the application of an electric field puts pressure on the material and the pressure on the material causes an electric field to be created as a result. In this way, these materials change their dimensions when an electric field is applied to them and when they are stressed by mechanical forces, the change in their dimensions generates linear electric charges. It should be noted that when the electric field is removed and/or there is no mechanical stress, these materials return to their original symmetrical state, without deformation. Figure 6 below shows the effect mentioned for quartz, the material which makes up the piezoelectric sensor used in this work. While Figure 7 shows examples of piezoelectric sensors with a quartz crystal core and coated with metal (bronze, aluminum or copper) like the one used in this work. (Tired Tires, 2016; Arnau Vives, 2008; R7 TECNOLOGIA, 2016)

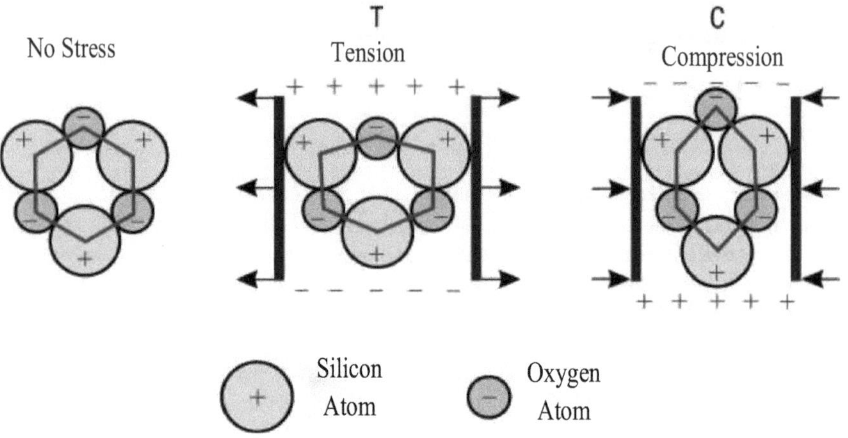

Figure 6: Piezoelectric effect in quartz.
Source: Reproduced and Adapted Piezoelectricity - Piezoelectric Effect in Quartz...(Tired Tires, 2016)

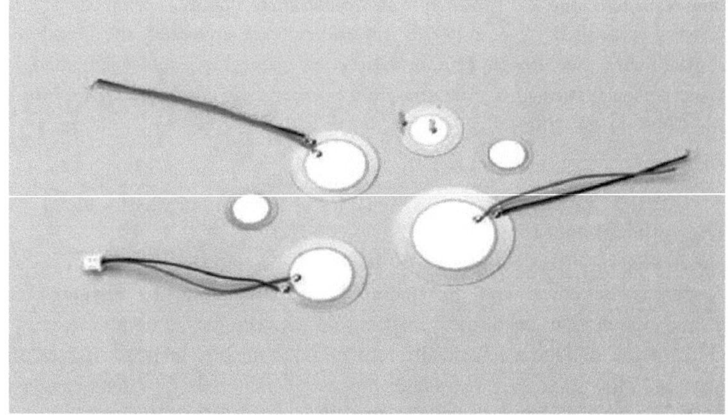

Figure 7: Piezoelectric sensors with quartz crystal core and metal coating (bronze, aluminum or copper).
Source: Reproduced and Adapted Piezoelectric Sensors... (R7 TECNOLOGIA, 2016)

3.2. State of the Art of Piezoelectricity

Piezoelectricity is currently used in a variety of applications in industry and cities. Examples include its extensive use in manufacturing processes for sensing purposes, acting as a load cell, machinery vibration meter, transducer to drive devices, etc. In addition, it is worth mentioning its use in energy generation within the urban infrastructure embedded in the sidewalks and roads of some cities. As an example, it is worth mentioning: - In Brazil and around the world, there is a commercial solution used to supply energy to the batteries of emergency telephones installed on roads as a carpet embedded in the ground; - Worldwide, some speed cameras also use piezoelectricity as sensors; - In the Netherlands, this technology is also installed in stretches of asphalt and supplies energy to the batteries responsible for LED lighting on the road itself; - In New York, piezoelectric sensors are installed in sections of the sidewalk in Times Square to generate energy for lighting; - In a metro station in Paris, piezoelectricity is built into the staircases where many people circulate in order to illuminate parts of the station itself. There are still few studies on the energy efficiency and generating capacity of this technology, but some cite that the generating capacity of a piezoelectric sensor, such as the one mentioned above and used in this experiment, with the step of a person is in the order of 0.0000164 kWh per step (KRISHNA and VIGNESH, 2015; HERRERA and RUMPF, 2014).

Chapter 4

4. METHODOLOGY

4.1. Requirements Analysis

In order to achieve the goal of building the desired prototype, it was necessary to use the following combination and configuration of hardware and software.

Given the hardware architecture of a microprocessor, in general terms, for the system to be intact, it needs to have basic elements in order for the processor to function. For the proposed system, a voltage of +5VDC is applied as the direct current power supply, a quartz crystal with a frequency of 8Mhz and the system reset is connected to +5VDC. Connected to the piezoelectric sensor is a loop with an operational amplifier in order to increase the DC levels of the sensor's response.

For the software, the PIC16F877 was programmed with the MikroC tool (MikroElektronika, 2016). This platform is used for programming in C for PICs, and the initial tests were carried out before the hardware was built using the Proteus simulator (Labcenter Electronics, 2016). To read the analog input of the PIC16F877, from 0 to 5VDC with 10 bits of resolution, the analog inputs contained in the PORTA bus are used. In order to get an adequate reading, a hardware interrupt based on an 8-bit TIMER0 with a time of 1ms is used. The interrupt complies with Equation 1 described below

$$FClk\_interrupção = \frac{FClk\_Processador}{4 * Pre\_Escala(256 - TMR0)} \quad (1)$$

With these parameters, it is necessary to configure the registers to behave as . Based on the tables provided by the PIC datasheet in question, illustrated in Figures 8 and 9, the following assignments for the registers are considered.

OPTION_REG:

Assigned Value **0x85h**

Bits:

7 - Value "1" - Disables pull-ups in PORTB.

6 - Value "0" - Interrupt on the falling edge of RB0 (external interrupt).

5 - Value "0" - Interrupt based on the PLL's internal clock.

4 - Value "0" - If external interruption increments on the rising edge of RA4. If internal interrupt increments on the rising edge of T0CKI.

3 - Value "0" - Pre-scale connected to the TIMER0 module.

2 - Value "1" - Sets valid Pre-scale value of 64.

1 - Value "0" " - Sets valid Pre-scale value of 64.

0 - Value "1" " - Sets valid Pre-scale value of 64.

REGISTER 2-2: OPTION_REG REGISTER (ADDRESS 81h, 181h)

R/W-1	R/W-1	R/W-1	R/W-1	R/W-1	R/W-1	R/W-1	R/W-1
$\overline{RBPU}$	INTEDG	T0CS	T0SE	PSA	PS2	PS1	PS0

bit 7 bit 0

bit 7 **$\overline{RBPU}$:** PORTB Pull-up Enable bit
1 = PORTB pull-ups are disabled
0 = PORTB pull-ups are enabled by individual port latch values

bit 6 **INTEDG**: Interrupt Edge Select bit
1 = Interrupt on rising edge of RB0/INT pin
0 = Interrupt on falling edge of RB0/INT pin

bit 5 **T0CS**: TMR0 Clock Source Select bit
1 = Transition on RA4/T0CKI pin
0 = Internal instruction cycle clock (CLKO)

bit 4 **T0SE**: TMR0 Source Edge Select bit
1 = Increment on high-to-low transition on RA4/T0CKI pin
0 = Increment on low-to-high transition on RA4/T0CKI pin

bit 3 **PSA**: Prescaler Assignment bit
1 = Prescaler is assigned to the WDT
0 = Prescaler is assigned to the Timer0 module

bit 2-0 **PS2:PS0**: Prescaler Rate Select bits

Bit Value	TMR0 Rate	WDT Rate
000	1 : 2	1 : 1
001	1 : 4	1 : 2
010	1 : 8	1 : 4
011	1 : 16	1 : 8
100	1 : 32	1 : 16
101	1 : 64	1 : 32
110	1 : 128	1 : 64
111	1 : 256	1 : 128

Legend:		
R = Readable bit	W = Writable bit	U = Unimplemented bit, read as '0'
- n = Value at POR	'1' = Bit is set	'0' = Bit is cleared x = Bit is unknown

Note: When using Low-Voltage ICSP Programming (LVP) and the pull-ups on PORTB are enabled, bit 3 in the TRISB register must be cleared to disable the pull-up on RB3 and ensure the proper operation of the device

Figure 8: PIC OPTION_REG register table, addresses from 81h to 181h.
Source: Reproduced and Adapted from REGISTER 2-2: OPTION_REG REGISTER (ADDRESS 81h, 181h)... (Microchip Technology, 2013

INTCON:

Assigned Value **0xA0h**

Bits:

7 - Value "1" - Enables all global interrupts.

6 - Value "0" - Disables interrupt peripherals.

5 - Value "1" - Enables all interrupt masks.

4 - Value "0" - Disables external interruption.

3 - Value "0" - Disables interrupt switching in PORTB.

2 - Value "0" - Interrupt overflow.

1 - Value "0" " - Does not enable interrupt on RB0.

0 - Value "0" " - Does not exchange RB7:RB4 state on interrupt.

REGISTER 2-3: INTCON REGISTER (ADDRESS OBh. 8Bh, 10Bh, 18Bh)

R/W-0 R/W-0 R/W-0 R/W-0 R/W-0 R/W-0 R/W-0 R/W-x

GIE I PEIE I TMR0IE| INTE | RBIE | TMR0IF | INTF RBIF

bit 7 bitO

bit 7 GIE: Global Interrupt Enable bit
1 = Enables all unmasked interrupts o = Disables all interrupts

bit 6 PEIE: Peripheral Interrupt Enable bit
1 = Enables all unmasked peripheral interrupts o = Disables all peripheral interrupts

bit 5 TMROIE: TMR0 Overflow Intenupt Enable bit 1 = Enables the TMR0 interrupt o = Disables the TMRO interrupt

bit 4 INTE: RBO/INT External Interrupt Enable bit 1 = Enables the RBO/INT external interrupt o = Disables the RBO/INT external interrupt

M3 RBIE: RB Port Change Intenupt Enable bit 1 = Enables the RB port change interrupt o = Disables the RB port change interrupt

bit 2 TMROIF: TMRO Overflow Interrupt Flag bit
1 = TMRO register has overflowed (must be cleared in software) o = TMRO register did not overflow

bit 1 INTF: RBO/INT External Interrupt Flag bit
1 = The RBO/INT external intenupt occuned (must be cleared in software) o = The RBO/INT external intenupt did not occur

bito RBIF: RB Port Change Interrupt Flag bit

1 = At least one of the RB7:RB4 pins changed state; a mismatch condition will continue to set

the bit. Reading PORTB will end the mismatch condition and allow the bit to be cleared

(must be cleared in software).

o = None of the RB7:RB4 pins have changed state

Legend:

R= Readable bit W= Writable bit U= Unimpleniented bit. read as 'O'

* n= Value at POR '1'= Bit is set 'O'= Bit is cleared x= Bit is unknown

Figure 9: PIC INTCON REGISTER register table, addresses 0Bh, 8Bh, 10Bh, 18Bh.

Source: Reproduced and Adapted from INTCON REGISTER (0Bh, 8Bh, 10Bh, 18Bh)... (Microchip Technology, 2013)

Equation 2 below shows the calculation for the 1ms interruption.

$$1000 = \frac{8000000}{4*64(256-TM?0)} \quad (2)$$

Then,

TMR0 = 224.75 for 1ms interruption, rounded up to 225.

Next, we need to consider the analog readout that needs to be implemented for this project. The PIC16F877 has analog input modules capable of reading a voltage of 0-5VDC with 10-bit resolution. For this process to take place, it is necessary to configure the ADCON0 and ADCON1 registers for this function as follows. To do this, it is also necessary to refer to the datasheet tables detailing them, shown in Figures 10 and 11 below.

ADCON0:

Assigned Value **0x81h**
Bits:
7 - Value "1" - Sets the conversion rate in Fosc/32.
6 - Value "0" - Sets conversion rate in Fosc/32.
5 - Value "0" - Selects channel zero for reading (multiplexed 8 channels).
4 - Value "0" - Selects channel zero for reading (multiplexed 8 channels).
3 - Value "0" - Selects channel zero for reading (multiplexed 8 channels).
2 - Value "0" - If '1' indicates conversion request if '0' conversion completed.
1 - Value "0" " - Not used.
0 - Value "1" " - Switches the A/D conversion module on.

REGISTER 11-1: ADCONO REGISTER (ADDRESS22 Fh)

R/W-0	R/W-0	R/W-0	R/W-0	R/W-0	R/W-0	U-0	R/W-0
ADCS1	ADCS0	CHS2	CHS1	CHS0	GO/DONE	—	ADON
bit 7							bit 0

bit 7-6 **ADCS1:ADCS0:** A/D Conversion Clock Select bits (ADCON0 bits in **bold**)

ADCON1 <ADCS2>	ADCON0 <ADCS1:ADCS0>	Clock Conversion
0	00	Fosc/2
0	01	Fosc/8
0	10	Fosc/32
0	11	FRC (clock derived from the internal A/D RC oscillator)
1	00	Fosc/4
1	01	Fosc/16
1	10	Fosc/64
1	11	FRC (clock derived from the internal A/D RC oscillator)

bit 5-3 **CHS2:CHS0:** Analog Channel Select bits

000 = Channel 0 (AN0)
001 = Channel 1 (AN1)
010 = Channel 2 (AN2)
011 = Channel 3 (AN3)
100 = Channel 4 (AN4)
101 = Channel 5 (AN5)
110 = Channel 6 (AN6)
111 = Channel 7 (AN7)

Note: The PIC16F873A/876A devices only implement A/D channels 0 through 4; the unimplemented selections are reserved. Do not select any unimplemented channels with these devices.

bit 2 **GO/DONE:** A/D Conversion Status bit

When ADON = 1:

1 = A/D conversion in progress (setting this bit starts the A/D conversion which is automatically cleared by hardware when the A/D conversion is complete)

0 = A/D conversion not in progress

bit 1 **Unimplemented:** Read as '0'

bit 0 **ADON:** A/D On bit

1 = A/D converter module is powered up

0 = A/D converter module is shut-off and consumes no operating current

Legend:		
R = Readable bit	W = Writable bit	U = Unimplemented bit, read as '0'
- n = Value at POR	'1' = Bit is set	'0' = Bit is cleared x = Bit is unknown

© 2003 Microchip Technology Inc. DS39582B-page 127

Figure 10: PIC ADCON0 REGISTER register table, address 1Fh.

Source: Reproduced and Adapted REGISTER 11-1: ADCON0 REGISTER (ADDRESS: 1Fh)...(Microchip Technology, 2013)

ADCON1:

Assigned Value **0x0Eh**

Bits:

7 - Value "0" - Value 10 bits right-justified in the ADRESH and ADRESL registers.

6 - Value "0" - Sets the sampling rate (clock conversion) to Fosc/2.

5 - Value "0" - Not used.

4 - Value "0" - Not used.

3 - Value "1" - All digital PORT, only analog bit 0.

2 - Value "1" - All digital PORT, only analog bit 0.

1 - Value "1" " - All digital PORT, only analog bit 0.

0 - Value "0" " - All digital PORT, only analog bit 0.

REGISTER 11-2: ADCON1 REGISTER (ADDRESS 9Fh)

R/W-0	R/W-0	U-0	U-0	R/W-0	R/W-0	R/W-0	R/W-0
ADFM	ADCS2	—	—	PCFG3	PCFG2	PCFG1	PCFG0

bit 7 — bit 0

bit 7 **ADFM:** A/D Result Format Select bit

1 = Right justified. Six (6) Most Significant bits of ADRESH are read as '0'.

0 = Left justified. Six (6) Least Significant bits of ADRESL are read as '0'.

bit 6 **ADCS2:** A/D Conversion Clock Select bit (ADCON1 bits in shaded area and in **bold**)

ADCON1 <ADCS2>	ADCON0 <ADCS1:ADCS0>	Clock Conversion
0	00	FOSC/2
0	01	FOSC/8
0	10	FOSC/32
0	11	FRC (clock derived from the internal A/D RC oscillator)
1	00	FOSC/4
1	01	FOSC/16
1	10	FOSC/64
1	11	FRC (clock derived from the internal A/D RC oscillator)

bit 5-4 **Unimplemented:** Read as '0'

bit 3-0 **PCFG3:PCFG0:** A/D Port Configuration Control bits

PCFG <3:0>	AN7	AN6	AN5	AN4	AN3	AN2	AN1	AN0	VREF+	VREF-	C/R
0000	A	A	A	A	A	A	A	A	VDD	VSS	8/0
0001	A	A	A	A	VREF+	A	A	A	AN3	VSS	7/1
0010	D	D	D	A	A	A	A	A	VDD	VSS	5/0
0011	D	D	D	A	VREF+	A	A	A	AN3	VSS	4/1
0100	D	D	D	D	A	D	A	A	VDD	VSS	3/0
0101	D	D	D	D	VREF+	D	A	A	AN3	VSS	2/1
011x	D	D	D	D	D	D	D	D	—	—	0/0
1000	A	A	A	A	VREF+	VREF-	A	A	AN3	AN2	6/2
1001	D	D	A	A	A	A	A	A	VDD	VSS	6/0
1010	D	D	A	A	VREF+	A	A	A	AN3	VSS	5/1
1011	D	D	A	A	VREF+	VREF-	A	A	AN3	AN2	4/2
1100	D	D	D	A	VREF+	VREF-	A	A	AN3	AN2	3/2
1101	D	D	D	D	VREF+	VREF-	A	A	AN3	AN2	2/2
1110	D	D	D	D	D	D	D	A	VDD	VSS	1/0
1111	D	D	D	D	VREF+	VREF-	D	A	AN3	AN2	1/2

A = Analog input D = Digital I/O

C/R = # of analog input channels/# of A/D voltage references

Legend:		
R = Readable bit	W = Writable bit	U = Unimplemented bit, read as '0'
- n = Value at POR	'1' = Bit is set	'0' = Bit is cleared x = Bit is unknown

Note:	On any device Reset, the port pins that are multiplexed with analog functions (ANx) are forced to be an analog input.

Figure 11: PIC ADCON1 REGISTER register table, address 9Fh.

Source: Reproduced and Adapted REGISTER 11-2: ADCON1 REGISTER (ADDRESS: 9Fh)...(Microchip Technology, 2013)

4.2. Development

The software developed in the C language codes an algorithm that always works linked to interrupt events. In other words, when a TMR0 interrupt occurs, the T0IF_bit bit goes to logic level 1. When this happens, the system interrupts the data stack and responds to the interrupt with priority. At this point a digital analog conversion is requested, when the GO/DONE bit of the ADCON0 register goes to zero it is ready for the conversion to be read. At this point, the values in ADRESH and ADREL, which are the analog input registers, are received in order to obtain a value allocated in an INT variable with a size of 16 bits. To do this, shift instructions are executed to adjust the values. With this process covered in a comparative chain of "if"s or "if"s, the bargraph is activated to display the amplitude of the signal generated by the sensor. Annex A shows the main sections identified in the commented code, in accordance with the logic established for this program, which was simulated and recorded in the hardware.

4.3. Simulation

Based on the code described above, a simulation was designed using Proteus software. This platform has a complete simulation of various hardware components, including the PIC16F877. In this way, the entire schematic for the board that was implemented was built and tested in advance using the software, recording the described code on the PIC and making all the necessary connections. In other words, simulating the complete operation of the physical project. Figure 12 below shows all these components in the simulator. The upper TRIMPOT_VARIATION functions as a potentiometer to adjust the gain of the amplifier to the parameters (ranges) described in the code, since the piezo signal may be too low for the microcontroller port to read. The second, lower TRIMPOT_VARIATION, plays the role of emulating the electric piezo sensor response for simulation purposes. RESET, meanwhile, has the function of returning the system to its initial settings, a trigger in the event of a problem with the circuit. The MICROCONTROLLER is the PIC16F877 with the code inserted and recorded, and is the heart of this system. It is connected to the BARGRAPH in order to display, from bottom to top, the pressure intensity with which the piezo's response is perceived.

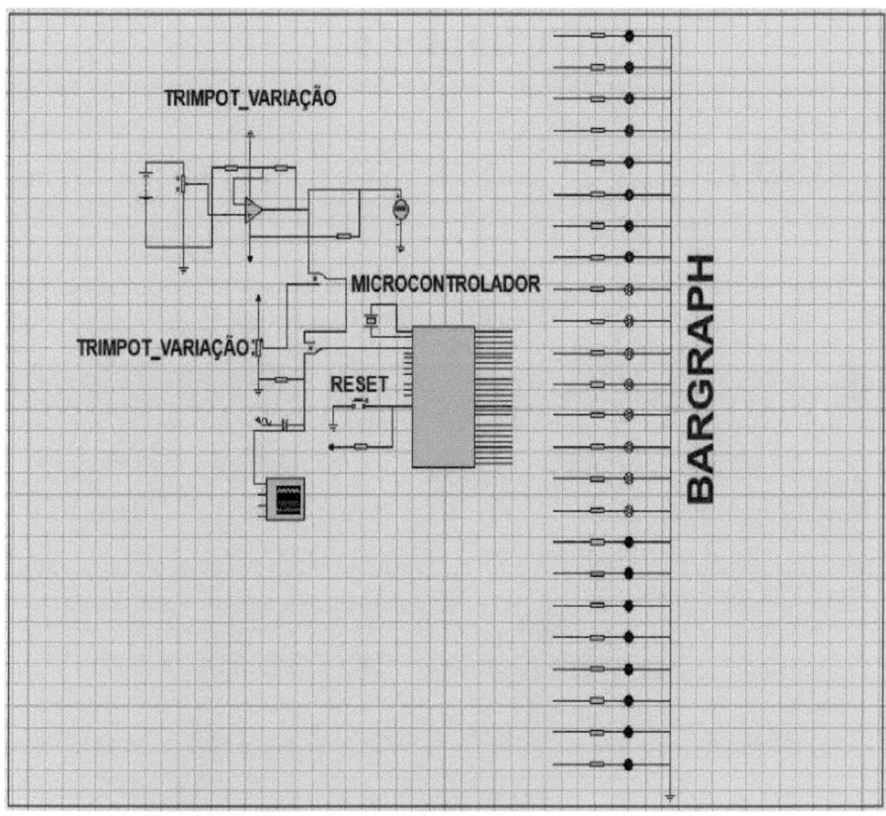

Figure 12: Simulation schematic via Proteus software.
Source: Own figure.

Chapter 5

5. RESULTS

The physical construction of the components resulted in the prototype shown in this chapter. Figure 13 shows a component for recording and debugging PICs, through which the code developed and simulated on the computer for the PIC was recorded (MicroGenios, 2016).

Figure 13: MicroICD ZIF Recorder - PIC Recorder and Debugger via USB 2.0.
Source: Own figure.

The phenolite board with all the components connected as planned in the project and simulated is shown in Figures 14 and 15 below. Each component is indicated: The PIC representing the embedded microprocessor; The Bargraph with a series of LEDs that light up from minimum to maximum levels to show the intensity ranges of the mechanical pressure variation suffered by the piezoelectric sensor; The electric piezo that captures the pressure variation through an electric charge; The Source that feeds the circuit; the Amplifier that is associated with the Trimpot to categorize the levels read from the sensor's response. Figure 15 shows the piezoelectric actuator and the response shown by the Bargraph on this board.

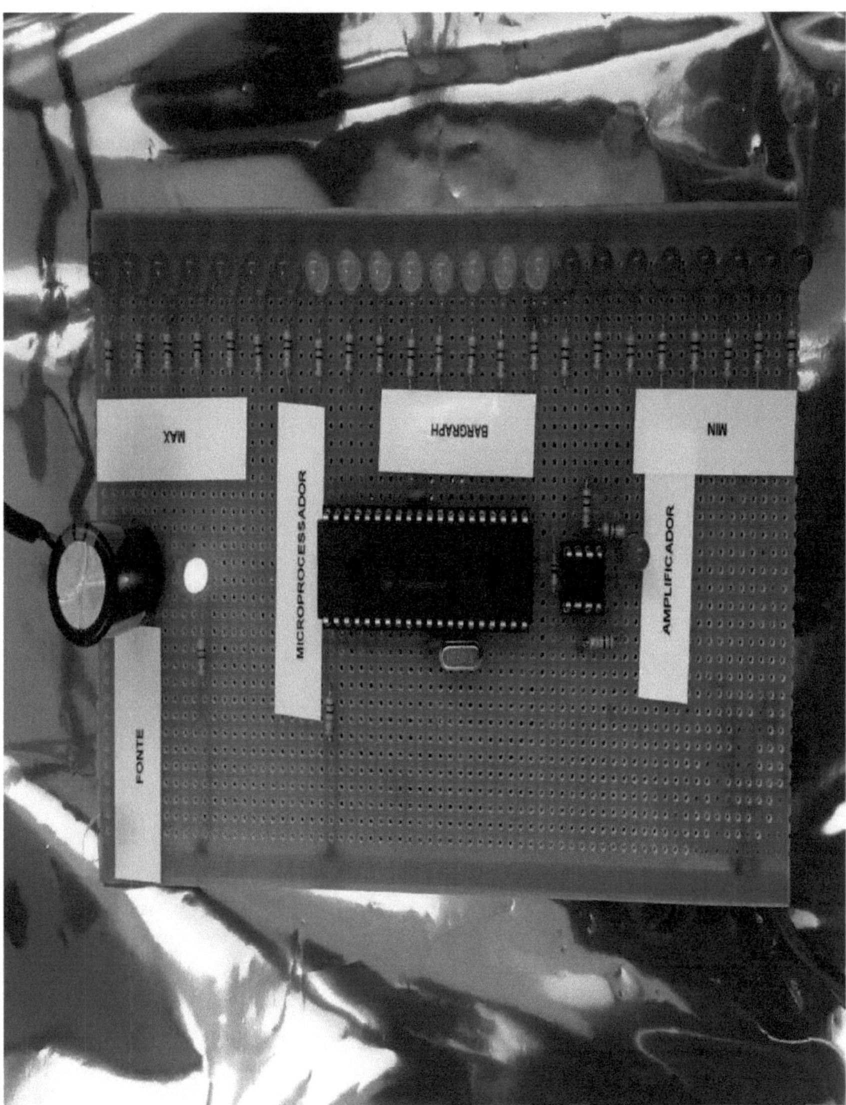

Figure 14: Phenolite board with all the components connected.
Source: Own figure.

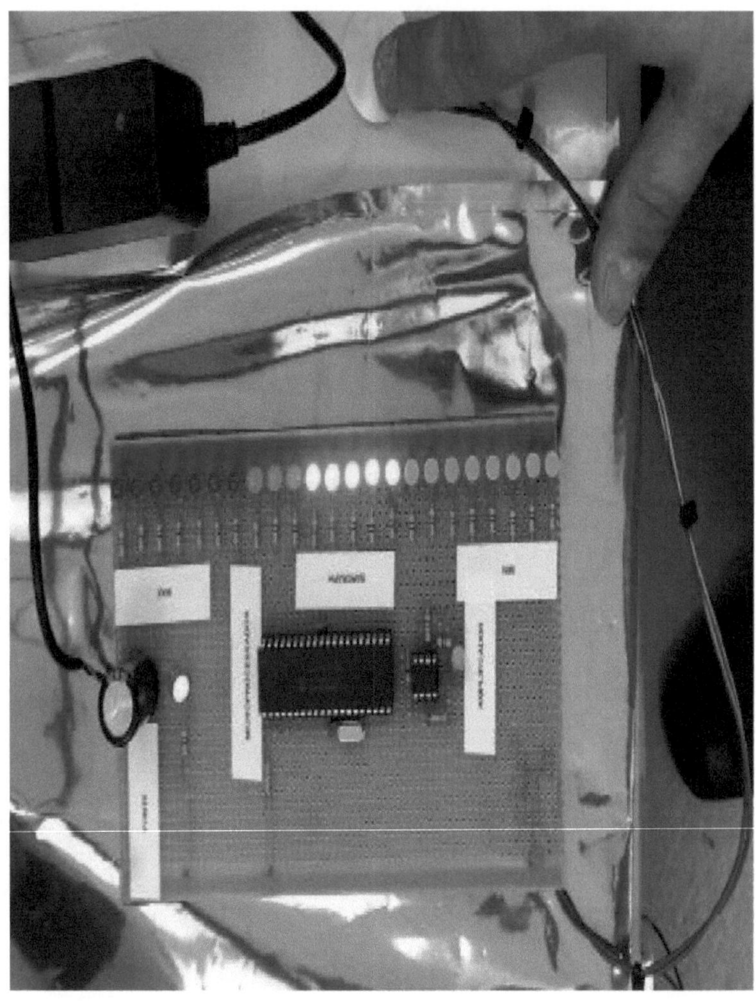

Figure 15: Developed board in operation with all components connected to the piezoelectric sensor.
Source: Own figure.

In addition, in order to demonstrate the applicability of the prototype in urban infrastructure systems, the sensor and board are attached to a model, as shown in Figure 16. This is a ferrorama, to simulate the use of this technology on a large scale. That is, producing energy from the losses of motor-kinetic engines within the context of cities.

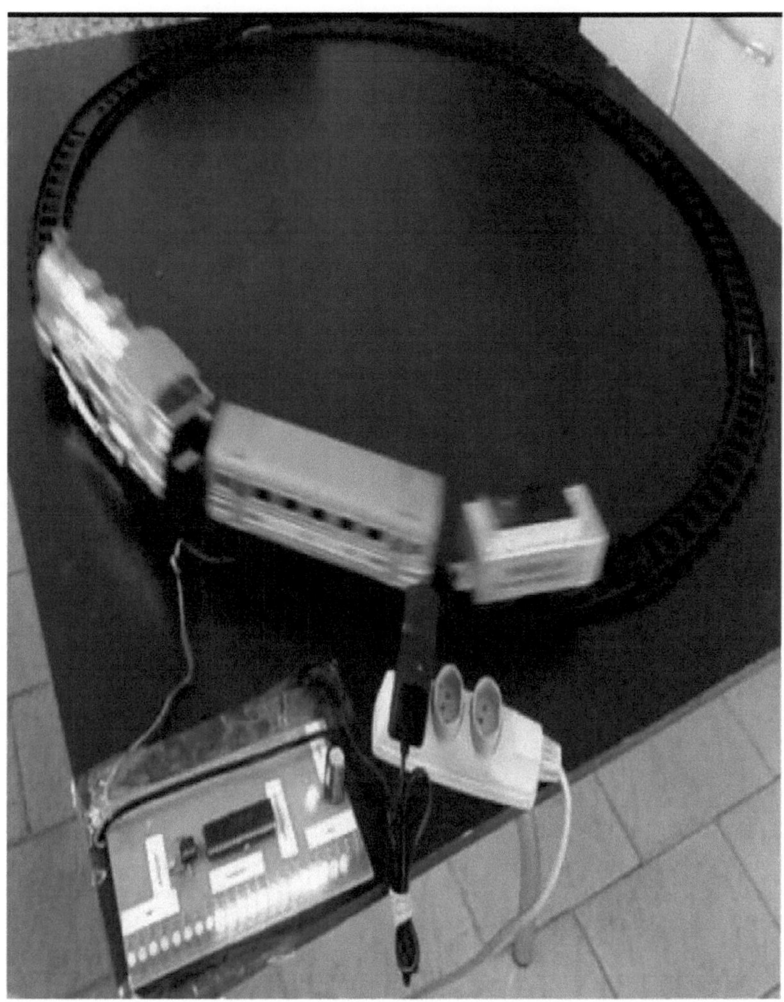

Figure 16: Ferrorama mock-up showing the automated demonstration of the experiment.

Source: Own figure

CONCLUSION

Given the search for renewable energy sources, as well as the concentration of modern life in the context of cities. One of the main advantages of applying piezoelectricity is that it uses the motor-kinetic energy wasted by means of transport and transforms it into electrical energy. This results in clean energy, as it is not dependent on natural resources such as fossil fuels. In addition, this solution can be installed without major construction work such as a hydroelectric plant or huge areas for installing equipment such as solar parks and wind farms. However, the main disadvantages of this technology are the variable generation, battery storage and distribution of the energy produced. This is because generation varies according to the flow of transport movements, the cost of investing in batteries is always high and energy distribution requires frequency inverters, since the energy obtained is in direct current. In most cases where this technology is implemented, the energy generated is spent immediately, which limits its applicability.

It is also worth considering that this technology for obtaining electricity using piezoelectric material is still in its infancy when considered on a large scale. This is because, although there is still not a wide range of information and studies on methods that use piezoelectric material to obtain energy, some experiments have already proven that it is possible to generate energy satisfactorily and charge batteries using these components. The analysis carried out during the tests for this project showed that there are many variables that influence the system's performance. Add to this the fact that each of these variables is extremely sensitive, and advanced engineering is always required to optimize the performance of such a circuit. This may be one of the reasons why this type of application is not yet so widespread. In addition, the system's own variable and unmeasurable performance means that the return on investment time is very long, increasing the risks.

Finally, it is clear that piezoelectric material already has several applications, but a lot of investment in development is still needed for its use as a transducer of mechanical energy into electrical energy to become viable at commercial levels. However, considering that the search for renewable energies is a trend, this material should become more popular in the near future.

With this objective in mind and considering the context of cities, we can see that the development of this technology points to three aspects as future studies. The first is the regulatory framework that needs to be debated and created in order to interconnect these systems with this technology to the utilities' electricity grid, given that it generates in direct current systems and with continuous generation variability, presenting a challenge that could harm the Electric Power System. The second is the very important involvement of traffic engineering companies in choosing the best locations to embed this technology. A multidisciplinary team is needed to combine studies in this area with knowledge of electrical engineering and electronics to make this solution viable. The third points to manufacturing, with a great need for industry to dedicate itself to building larger-scale solutions, moving away from simple construction of small transducers to robust and effective generation devices that can rival other alternative energy sources.

REFERENCES

Arnau Vives, 2008. Piezoelectric Transducers and Applications 2nd Edition 2008. Berlin, Germany. Publisher Springer-Verlag Berlin Heidelberg. 532p. ISBN: 9783540775089.

Canadian Institute of Actuaries, 2015. Canadian Institute of Actuaries. CIA Climate Change and Sustainability Forum. Available at: <http://www.cia- ica.ca/professional-development/meetings/meeting-archives/2015/cia-climate- change-and-sustainability-forum/>. Accessed on: April, 20, 2016.

CIA, 2011. Central Intelligence Agency. CIA Highlights Sustainability and Conservation Initiatives on Earth Day. Available at: <https://www.cia.gov/news-information/press-releases-statements/press-release-2011/earth-day.html>. Accessed on: April, 20, 2016.

CIA, 2016. Central Intelligence Agency. The World Fact Book. Available at: < https://www.cia.gov/library/publications/the-world-factbook/geos/xx.html>. Accessed on: April, 20, 2016.

Costa, Cesar Da, Mesquita, Leonardo and Pinheiro, Eduardo; 2011. Elements Of Programmable Logic With Vhdl And Dsp: Theory And Practice 1st Edition 2011. São Paulo-SP, Brazil. Editora Érica. 292p. ISBN: 9788536503127 - 8536503122.

Cotta, Geovanna; 2007. Unicamp - State University of Campinas. Piezoelectric effect: experiments. Diponível em: <http://www.ifi.unicamp.br/~lunazzi/F530_F590_F690_F809_F895/F809/F809_sem1_2007/GeovannaL_Cotta_RF1.p df>. Accessed on: April, 20, 2016.

Ecotricity, 2016. Ecotricity Inc. The End Of Fossil Fuels. Available at: <https://www.ecotricity.co.uk/our-green-energy/energy-independence/the-end-of-fossil-fuels>. Accessed on: April, 20, 2016.

HERRERA, JOSE ADRIAN and RUMPF, RAYMOND C., 2014. Piezoelectric Technology. Influenced by Electromagnetic Induction. COLLEGE OF ENGINEERING - THE UNIVERSITY OF TEXAS AT EL PASO.Available at: <http://emlab.utep.edu/ee4347appliedem /projs_F2014/ Piezoelectric%20Technology.pdf>. Accessed on: April, 20, 2016.

IBGE, 2010. Brazilian Institute of Geography and Statistics. Synopsis of the 2010 Demographic Census - Brazil. Available at: <http://www.censo2010.ibge.gov.br/sinopse/index.php?dados=8>. Accessed on: April, 20, 2016.

ICNT, 2013. National Institute of Science and Technology - OBSERVATORY OF METROPOLES. EVOLUTION OF THE AUTOMOBILE AND MOTORCYCLE FLEET IN BRAZIL 2001 - 2012 (Report 2013). Available at: <http://www.observatoriodasmetropoles.net/download/auto_motos2013.pdf>. Accessed on: April, 20, 2016.

IMF, 2011. INTERNATIONAL MONETARY FUND. OIL SCARCITY, GROWTH, AND GLOBAL IMBALANCES. WORLD ECONOMIC OUTLOOK: TENSIONS FROM THE TWO-SPEED RECOVERY, CHAPTER 3, April 2010, pg. 89-122. Available at: <http://www.imf.org/external/pubs/ft/weo/2011/01/pdf/c3.pdf> Accessed on: April, 20, 2016.

KRISHNA, S. and VIGNESH, S., 2015. DESIGN OF ENERGY CAPTURING MEDIUM USING PIEZOELECTRIC EFFECT. International Journal of Scientific Engineering and Applied Science (IJSEAS) - Volume-1, Issue-4, July 2015. ISSN: 2395-3470.Available at: <http://ijseas.com/volume1/ v1i4/ijseas20150419.pdf>. Accessed on: April, 20, 2016.

Labcenter Electronics, 2016. Labcenter Electronics Inc. Proteus Design Suite. Available at: <http://www.labcenter.com/index.cfm>. Accessed on: April, 20, 2016.

Microchip Technology, 2013. Microchip Technology Inc. PIC16F87X - 28/40-Pin 8-Bit CMOS FLASH Microcontrollers. Available at: <http://ww1.microchip.com/ downloads/en/DeviceDoc/30292D.pdf>. Accessed on: April, 20, 2016.

MicroGenios, 2016. MicroGenios. MicroICD ZIF Recorder - PIC and dsPIC Recorder and Debugger via USB 2.0. Available at: <http://www.microgenios.com/?1.32.0.0,353, gravador-microicd-zif-gravador-e- debugador-de-pic-e-dspic-via-usb-2.0.html>. Accessed on: April, 20, 2016.

MikroElektronika, 2016. MikroElektronika Inc. mikroC PRO for PIC - Your PIC best friend. Available at: <http://www.mikroe.com/mikroc/pic/>. Accessed on: April, 20, 2016.

Miyadaira, Alberto Noboru, 2009. Pic18 Microcontrollers - Learn and Program in C Language 1st Edition 2009. São Paulo-SP, Brazil. Editora Érica. 268p. ISBN: 9788536502441.

OLIVEIRA, Paulo André de and SIMON, Elias José, 2004. Productive electricity consumption and the value of agricultural production in the Botucatu region. In: ENCONTRO DE ENERGIA NO MEIO RURAL, 5., 2004, Campinas. Proceedings online... Available at: <http://www.proceedings.scielo.br/scielo.php? script=sci_arttext&pid=MSC0000000022004000200024&lng=en&nrm=abn>. Accessed on: April, 20, 2016.

R7 TECNOLOGIA, 2016. Hardware Community - R7 TECNOLOGIA. Piezoelectric sensors. Available at: <http://www.hardware.com.br/comunidade /sensores-piezoeletricos/984664/>. Accessed on: April, 20, 2016.

Souza, David Jose de, 2003. Desbravando o Pic - Ampliado e Atualizado para Pic16f628a - 6ª Edition 2003. São Paulo-SP, Brazil. Editora Érica. 268p. ISBN: 8571948674.

Statista, 2016 A. Statista Inc. Worldwide electricity generation from 1990 to 2013 (in terawatt hours). Available at: <http://www.statista.com/statistics/ 270281/electricity-

generation-worldwide/>. Accessed on: April, 20, 2016.

Statista, 2016 B. Statista Inc. Average electricity prices globally in 2015, by select country (in U.S. dollars per kilowatt hour). Available at: <http://www.statista.com/statistics/477995/global-prices-of-electricity-by-select-country/>. Accessed on: April, 20, 2016.

Tired Tires, 2016. Tired Tires. Piezoelectricity - Piezoelectric Effect in Quartz. Available at: <http://dev.nsta.org/evwebs/2014102/news/default.html>. Accessed on: April, 20, 2016.

United Nations, 2014. United Nations. World's population increasingly urban with more than half living in urban areas. Available at: <http://www.un.org/en /development/desa/news/population/world-urbanization-prospects-2014.html>. Accessed on: April, 20, 2016.

WHO, 2014. World Health Organization. Global Health Observatory (GHO) data - Urban population growth. Available at: <http://www.who.int/gho/urban_health /situation_trends/urban_population_growth_text/en/>. Accessed on: April, 20, 2016

ANNEX A

Código comentado.

```
// Rotina sensor PiezoElétrico com Bargraph.
// Configura variáveis
bit teste1ms;
bit tempo100ms;
bit keyconversao;
unsigned int conversao;
unsigned char conversaoLSB;
unsigned char aux0, aux1;
unsigned char counterdelayconversao;
unsigned int counter100ms;
unsigned int controle;
// Interrupção 1ms
void interrupt()
{
 if (T0IF_bit)
 {
  T0IF_bit = 0;      // clear TMR0IF
  TMR0  = 225;
  counter100ms++;
  if(counter100ms > 0) // Controla o tempo de histerese
  {
   counter100ms = 0;
   tempo100ms = ~tempo100ms;
  }
  teste1ms = ~teste1ms;
    // Conversão A/D
  // Leitura AD
  if (counterdelayconversao==0)ADCON0 = ADCON0|0x04; // Pede conversão
  if((teste1ms)&&(keyconversao==0))
  {
   keyconversao = 1;
   counterdelayconversao++;
  }
  if((teste1ms==0)&&(keyconversao==1))keyconversao = 0;
  if (counterdelayconversao > 4)
  {
   if(!(ADCON0&0x04))
   {
    conversao = ADRESH<<2;
    conversaoLSB = ADRESL>>6;
    conversao = conversao + conversaoLSB;
   }
   counterdelayconversao = 0;
  }
 }
}
void main()
```

```
{
// Configura Interrupção
  OPTION_REG = 0x85;      // Assign prescaler to TMR0
  TMR0  = 225;            // Timer0 initial value
  INTCON = 0xA0;          // Enable TMRO interrupt
  PIE1 = 0x03;            // Habilita interrupções - somente Timer2 - PWM
// Configuração Ports
  TRISB = 0x00;           // Portb como saídas
  PORTB = 0x00;           // Inicializa sinais em zero
  TRISD = 0x03;           // Portd como entradas
  PORTD = 0x00;           // Inicializa sinais em zero
  TRISE = 0x00;           // Libera Portd para ser de uso geral
// Configuração Ports
  TRISC = 0x00;
  PORTC = 0x00;
  TRISD = 0x00;
  PORTD = 0x00;
// Inicializa Conversor A/D
  ADCON0 = 0x81; // Canal zero
  ADCON1 = 0x0E; // Usa referencia interna
// Inicializa variáveis
  teste1ms = 0;
  keyconversao = 0;
  conversao = 0;
  counterdelayconversao = 0;
  counter100ms = 0;
  tempo100ms = 0;
  conversaoLSB = 0;
  aux0 = 0;
  aux1 = 0;
  controle = 0;
// Loop Principal
  while(1)
  {
 /*
   aux0 = conversao>>8;
   aux1 = conversao&0xFF;
   PORTD = aux1;
   PORTC = aux0;
 */
   controle = conversao;
   // D7
   if (controle>41)
   {
    PORTD.RD7 = 1;
   }
   else
   {
    PORTD.RD7 = 0;
   }
```

```
// D6
if (controle>83)
{
 PORTD.RD6 = 1;
}
else
{
 PORTD.RD6 = 0;
}
// D5
if (controle>125)
{
 PORTD.RD5 = 1;
}
else
{
 PORTD.RD5 = 0;
}
// D4
if (controle>167)
{
 PORTD.RD4 = 1;
}
else
{
 PORTD.RD4 = 0;
}
// D3
if (controle>209)
{
 PORTD.RD3 = 1;
}
else
{
 PORTD.RD3 = 0;
}
// D2
if (controle>251)
{
 PORTD.RD2 = 1;
}
else
{
 PORTD.RD2 = 0;
}
// D1
if (controle>293)
{
 PORTD.RD1 = 1;
}
```

```
else
{
 PORTD.RD1 = 0;
}
// D0
if (controle>335)
{
 PORTD.RD0 = 1;
}
else
{
 PORTD.RD0 = 0;
}

// C7
if (controle>377)
{
 PORTC.RC7 = 1;
}
else
{
 PORTC.RC7 = 0;
}
// C6
if (controle>419)
{
 PORTC.RC6 = 1;
}
else
{
 PORTC.RC6 = 0;
}
// C5
if (controle>461)
{
 PORTC.RC5 = 1;
}
else
{
 PORTC.RC5 = 0;
}
// C4
if (controle>504)
{
 PORTC.RC4 = 1;
}
else
{
 PORTC.RC4 = 0;
}
```

```
// C3
if (controle>545)
{
 PORTC.RC3 = 1;
}
else
{
 PORTC.RC3 = 0;
}
// C2
if (controle>587)
{
 PORTC.RC2 = 1;
}
else
{
 PORTC.RC2 = 0;
}
// C1
if (controle>629)
{
 PORTC.RC1 = 1;
}
else
{
 PORTC.RC1 = 0;
}
// C0
if (controle>671)
{
 PORTC.RC0 = 1;
}
else
{
 PORTC.RC0 = 0;
}

// B7
if (controle>713)
{
 PORTB.RB7 = 1;
}
else
{
 PORTB.RB7 = 0;
}
// B6
if (controle>755)
{
 PORTB.RB6 = 1;
```

```
}
else
{
 PORTB.RB6 = 0;
}
// B5
if (controle>797)
{
 PORTB.RB5 = 1;
}
else
{
 PORTB.RB5 = 0;
}
// B4
if (controle>839)
{
 PORTB.RB4 = 1;
}
else
{
 PORTB.RB4 = 0;
}
// B3
if (controle>881)
{
 PORTB.RB3 = 1;
}
else
{
 PORTB.RB3 = 0;
}
// B2
if (controle>923)
{
 PORTB.RB2 = 1;
}
else
{
 PORTB.RB2 = 0;
}
// B1
if (controle>965)
{
 PORTB.RB1 = 1;
}
else
{
 PORTB.RB1 = 0;
}
```

```
// B0
if (controle>1007)
{
 PORTB.RB0 = 1;
}
else
{
 PORTB.RB0 = 0;
}

}
}
```

Printed by Books on Demand GmbH, Norderstedt / Germany